Countries of the World

Ethiopia

by Muriel L. Dubois

Consultant:
Azeb Tadesse Lemma
UCLA James S. Coleman African Studies Center

Bridgestone Books
an imprint of Capstone Press
Mankato, Minnesota

Bridgestone Books are published by Capstone Press
151 Good Counsel Drive, P.O. Box 669, Mankato, Minnesota 56002
http://www.capstone-press.com

Library of Congress Cataloging-in-Publication Data
Dubois, Muriel L.
    Ethiopia/by Muriel L. Dubois.
    p. cm.—(Countries of the world)
    Includes bibliographical references (p. 24) and index.
    ISBN 0-7368-0813-2
    1. Ethiopia—Juvenile literature. [1. Ethiopia.] I. Title. II. Countries of the world
(Mankato, Minn.)
DT373 .D83 2001
963—dc21   00-009668

Summary: Introduces the geography, animals, food, and culture of Ethiopia.

Editorial Credits

Tom Adamson, editor; Karen Risch, product planning editor; Linda Clavel, production designer
    and illustrator; Katy Kudela, photo researcher

Photo Credits

Michele Burgess, cover, 8, 10
Reuters/Wolfgang Rattay/Archive Photos, 18
Robert Maust/Photo Agora, 12
StockHaus Limited, 5 (top)
TRIP/A. Gasson, 14
Unicorn Stock Photos/Robert E. Barber, 16
Victor Englebert, 6, 20

Table of Contents

Fast Facts

Name: Federal Democratic Republic of Ethiopia

Capital: Addis Ababa

Population: More than 64 million

Official Language: Amharic

Religions: Christianity, Islam

Size: 435,184 square miles (1,127,127 square kilometers)

Ethiopia is about the size of the U.S. states of Texas, Oklahoma, and New Mexico combined.

Crops: Coffee, teff, wheat, barley

Maps

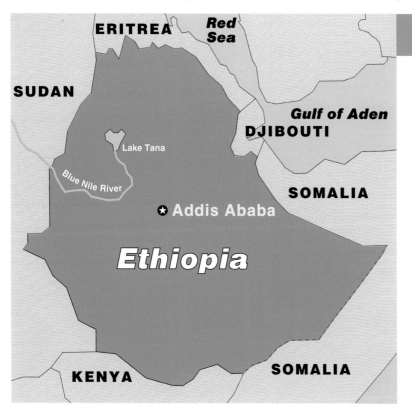

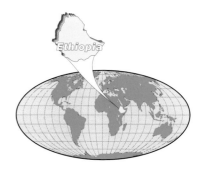

4

Flag

Ethiopia's flag has three stripes. The green stripe stands for rich lands and hope. The yellow stripe means religious freedom. The red stripe honors those who died for Ethiopia. A star on a blue background lies in the center of the flag. The color blue represents peace. The rays of the star stand for equality. Ethiopia adopted this flag in 1996.

Currency

The unit of currency in Ethiopia is the birr. One hundred Ethiopian cents equal one birr.

In 2000, about 8 birr equaled 1 U.S. dollar. About 5.5 birr equaled 1 Canadian dollar.

The Land

Ethiopia is a country in eastern Africa. Eritrea borders Ethiopia to the north. Djibouti and Somalia form the eastern border. Kenya lies to the south. Sudan is west of Ethiopia.

Mountains cover about two-thirds of Ethiopia. Most of these highlands run through the middle of the country. Lowlands surround the highlands. Ethiopia also has thick forests, grasslands, and deserts.

Lake Tana is Ethiopia's largest lake. It covers about 1,400 square miles (3,600 square kilometers). The Blue Nile River begins at Lake Tana. The Blue Nile River joins the White Nile River in Sudan to form the Nile River.

Ethiopia is near the equator. Mountain air keeps most of Ethiopia from becoming too hot. The rainy season lasts from the middle of June until early September. But some parts of the country receive little rain.

Ethiopians fish from reed boats on Lake Tana.

Life at Home

In large cities, some Ethiopians live like people in North America. They live in houses or apartments.

In the country, some families live in small round houses called gochos. People build the walls with saplings or bamboo stalks. They cover the walls with clay mud mixed with straw. They make the roof from grass. The gocho has one door. Some gochos have no windows.

Other families live in box-shaped houses. These houses have tin roofs that catch rain. Ethiopians do not want to waste any water.

Many Ethiopians are farmers. They grow only enough food for their families. Some families add a small room onto their house for their farm animals. They bring the animals indoors at night to keep them safe from hyenas. These wild dogs sometimes attack livestock.

In the country, some Ethiopians live in gochos.

Going to School

Not all children in Ethiopia go to school. Many children stay home to help care for farm animals or help cook meals. Families sometimes send only one child to school, usually a boy.

Some children attend Muslim schools or Christian schools. Others go to public schools. Only one-third of the students stay in elementary school for all eight years. Few students finish high school.

Many languages are spoken in Ethiopia. In school, children learn the language of their region. They also may study English and Amharic. Amharic is an official language in Ethiopia. Students also study math, science, history, and geography.

Many children and adults in Ethiopia cannot read or write. Ethiopia's government is working to help more children go to school.

Children begin school at age 7.

Ethiopian Food

Ethiopian farmers grow many grains including wheat, barley, corn, sorghum, and millet. A grain called teff is grown only in Ethiopia. Teff is used to make a bread called injera (in-JAIR-uh).

Families eat injera with a stew called wat. Wat is thick and hot. It is made with either vegetables or meat. A kind of hot pepper called berbere (ber-BER-ray) makes the stew spicy.

Religious customs affect what Ethiopians eat. Muslims and Ethiopian Orthodox Christians do not eat pork. Christians do not eat meat or dairy foods on Wednesdays and Fridays. During the holy month of Ramadan, Muslims fast from sunrise to sunset.

Children usually eat before their parents. Adults share one plate. They take food with their right hand only. Ethiopians believe it is not polite to eat with their left hand.

A stew called wat usually is eaten with injera.

Clothing

The clothing Ethiopians wear varies. Some people wear the same kinds of clothes that people in North America wear. Ethiopians sometimes wear traditional clothing.

One common type of clothing in Ethiopia is the shamma. Both men and women wear this rectangular-shaped shawl. Women wear the shamma over a dress. Men wear the shamma over a white shirt and pants.

Women sometimes wear a shash. They wind this colorful scarf around their heads. It usually covers all of a woman's hair.

On holidays, Ethiopians wear white cotton clothing. They decorate their clothing by sewing designs with bright thread. Some people add silk borders. Women sometimes wear a matching shamma on their heads.

Some women wear a colorful scarf called a shash.

Animals

Ethiopia has many kinds of wildlife. The Simien fox, the mountain nyala, and the walia ibex live only in Ethiopia. The Simien fox also is called the Ethiopian wolf. The mountain nyala is a kind of large antelope. It has long, curved horns. The walia ibex is a mountain goat.

Colobus monkeys live in mountain forests. Laws protect this animal. Hunters once killed the colobus monkey for its beautiful black and white fur.

Wild pigs, porcupines, and baboons live in the highlands. Some farmers must guard their family's crops from baboons. Troops of baboons will try to steal crops.

Hippopotamuses, crocodiles, and pythons live near rivers. Rhinos, lions, elephants, giraffes, zebras, gazelles, and buffalo live in protected parks and forests.

Colobus monkeys spend most of their time in trees.

Haile Gebrselassie set an Olympic record in 1996.

Sports and Games

Soccer is a popular sport in Ethiopia. Italian soldiers brought the game to Ethiopia in the 1930s. Many people who live in cities join soccer clubs. Each year, the best teams compete in the national championship tournament.

Students join high school teams. They play volleyball, basketball, and soccer. Some students also compete in gymnastics. The city of Addis Ababa hosts a sports festival every two years.

Ethiopia is famous for its fast runners. Ethiopian runners have won many gold medals in the Olympics. Adebe Bikila, a palace guard, won the marathon in the 1960 Olympics in Rome. He ran without shoes.

Two Ethiopians won gold medals in the 1996 Olympics in Atlanta. Fatuma Roba was the women's marathon winner. Haile Gebrselassie won the 10,000-meter race. He won this race again in the 2000 Olympics in Sydney.

Holidays and Celebrations

Ethiopia follows a 13-month calendar. Twelve of the months have 30 days. The 13th month, Pagame, has only five or six days. Each month has its own celebrations. Many Ethiopian holidays celebrate religious events.

Ethiopian Orthodox Christians celebrate Christmas on January 7. Ethiopians call Christmas ganna. Families begin to feast at midnight. Twelve days later, people celebrate Timkat. This day celebrates the baptism of Jesus Christ. A baptism is a ceremony that shows that someone has become a Christian. During Timkat, babies are baptized and children receive new clothes.

Muslims observe Ramadan. This month-long celebration honors the time when Muhammad was given the Quran. The Quran is the holy book of the religion of Islam. Muslims fast from sunrise to sunset during Ramadan.

Ethiopian priests dance during the Timkat celebration.

Hands On: Play Keliblibosh

Keliblibosh is a game you can play alone or with others. The game of jacks comes from Keliblibosh.

What You Need
10 or 12 pebbles

What You Do
1. Using one hand, gently scatter the pebbles on the ground.
2. Pick up one pebble and toss it into the air.
3. Before the pebble lands on the ground, pick up one of the scattered pebbles.
4. Put the pebble you picked up to one side.
5. Toss the first pebble in the air again.
6. This time, try to pick up two pebbles before it lands.
7. Repeat, trying to pick up an extra pebble each time.
8. When you miss, it is someone else's turn.

Learn to Speak Amharic

Many languages are spoken in Ethiopia. Amharic is an official language. It is used on TV and on the radio. The Amharic alphabet has 33 characters.

hello	ሰላም	selam	(suh-lahm)
father	አባት	abbat	(ahb-baht)
mother	እናት	enat	(eh-naht)
cat	ድመት	demet	(deh-meht)
dog	ውሻ	wusha	(woo-shah)

Words to Know

baptism (BAP-tiz-uhm)—a ceremony to show that someone is a Christian; the ceremony usually involves pouring water on the person's head.

Christian (KRISS-chuhn)—a person who follows the religion of Christianity; Christianity is based on the teachings of Jesus.

equator (i-KWAY-tur)—an imaginary line around the middle of Earth; regions near the equator are usually warm and wet.

fast (FAST)—to give up eating for a certain amount of time

marathon (MAR-uh-thon)—a race that covers 26 miles, 385 yards (about 42 kilometers)

Muslim (MUHZ-luhm)—a person who follows the religion of Islam; Islam is based on the teachings of Muhammad.

sapling (SAP-ling)—a young tree

troop (TROOP)—a group of baboons

Read More

Berg, Elizabeth. *Ethiopia*. Festivals of the World. Milwaukee: Gareth Stevens, 1999.

Waterlow, Julia. *A Family from Ethiopia*. Families around the World. Austin, Texas: Raintree Steck-Vaughn, 1998.

Useful Addresses and Internet Sites

Embassy of Ethiopia
3506 International Drive NW
Washington, DC 20009

Embassy of Ethiopia
#210-151 Slater Street
Ottawa, ON K1P 5H3
Canada

Embassy of Ethiopia
http://www.ethiopianembassy.org
The World Factbook 2000—Ethiopia
http://www.cia.gov/cia/publications/factbook/geos/et.html

Index